PLANT POWER

HEALING PLANTS

by Karen Latchana Kenney

pogo

Ideas for Parents and Teachers

Pogo Books let children practice reading informational text while introducing them to nonfiction features such as headings, labels, sidebars, maps, and diagrams, as well as a table of contents, glossary, and index.

Carefully leveled text with a strong photo match offers early fluent readers the support they need to succeed.

Before Reading

- "Walk" through the book and point out the various nonfiction features. Ask the student what purpose each feature serves.
- Look at the glossary together. Read and discuss the words.

Read the Book

- Have the child read the book independently.
- Invite him or her to list questions that arise from reading.

After Reading

- Discuss the child's questions. Talk about how he or she might find answers to those questions.
- Prompt the child to think more. Ask: Did you know about healing plants before reading this book? What more would you like to learn about them?

Pogo Books are published by Jump!
5357 Penn Avenue South
Minneapolis, MN 55419
www.jumplibrary.com

Library of Congress Cataloging-in-Publication Data

Names: Kenney, Karen Latchana, author.
Title: Healing plants / by Karen Latchana Kenney.
Description: Minneapolis, MN : Jump!, Inc., [2018]
Series: Plant power | Audience: Age 7-10. "Pogo Books."
Includes bibliographical references and index.
Identifiers: LCCN 2018003892 (print)
LCCN 2018008511 (ebook)
ISBN 9781624968761 (ebook)
ISBN 9781624968747 (hardcover : alk. paper)
ISBN 9781624968754 (paperback)
Subjects: LCSH: Plants–Therapeutic use–Juvenile literature.
Classification: LCC RS164 (ebook)
LCC RS164 .K425 2018 (print) | DDC 615.8/515–dc23
LC record available at https://lccn.loc.gov/2018003892

Editor: Jenna Trnka
Book Designer: Molly Ballanger

Photo Credits: guy42/Shutterstock, cover; Diana Taliun/Shutterstock, 1; fotohunter/Shutterstock, 3; Subbotina Anna/Shutterstock, 4; Evgeny Karandaev/Shutterstock, 5; Bucha Natallia/Shutterstock, 6-7; hanapon1002/iStock, 8-9; Photology1971/Shutterstock, 10; Trum Ronnarong/Shutterstock, 11; Julio Etchart/robertharding/Getty, 12-13; Izf/Shutterstock, 14-15; Zerbor/Shutterstock, 16; fizkes/Shutterstock, 17; blickwinkel/Alamy, 18-19; Yuriiyt/Shutterstock, 20-21; Nataly Studio/Shutterstock, 23.

Printed in the United States of America at Corporate Graphics in North Mankato, Minnesota.

TABLE OF CONTENTS

CHAPTER 1

FLOWERS AND FRUITS

Pretty yellow and white flowers dot a field. They look ordinary. But these flowers have healing powers. People have used them as **medicine** for hundreds of years.

This plant is chamomile. Its flowers are made into tea. Drinking it soothes stomach pains and can help you sleep. It is one of many plants that heal.

What is this pretty purple flower? Lavender! It is made into a sweet-smelling oil. It heals in many ways. Smelling it calms you. Rubbing it on your skin helps, too. It makes bug bites less itchy.

DID YOU KNOW?

Ancient Egyptians used lavender oil. They soaked cloth in it. Then they wrapped **mummies** in the cloth.

lavender

oranges

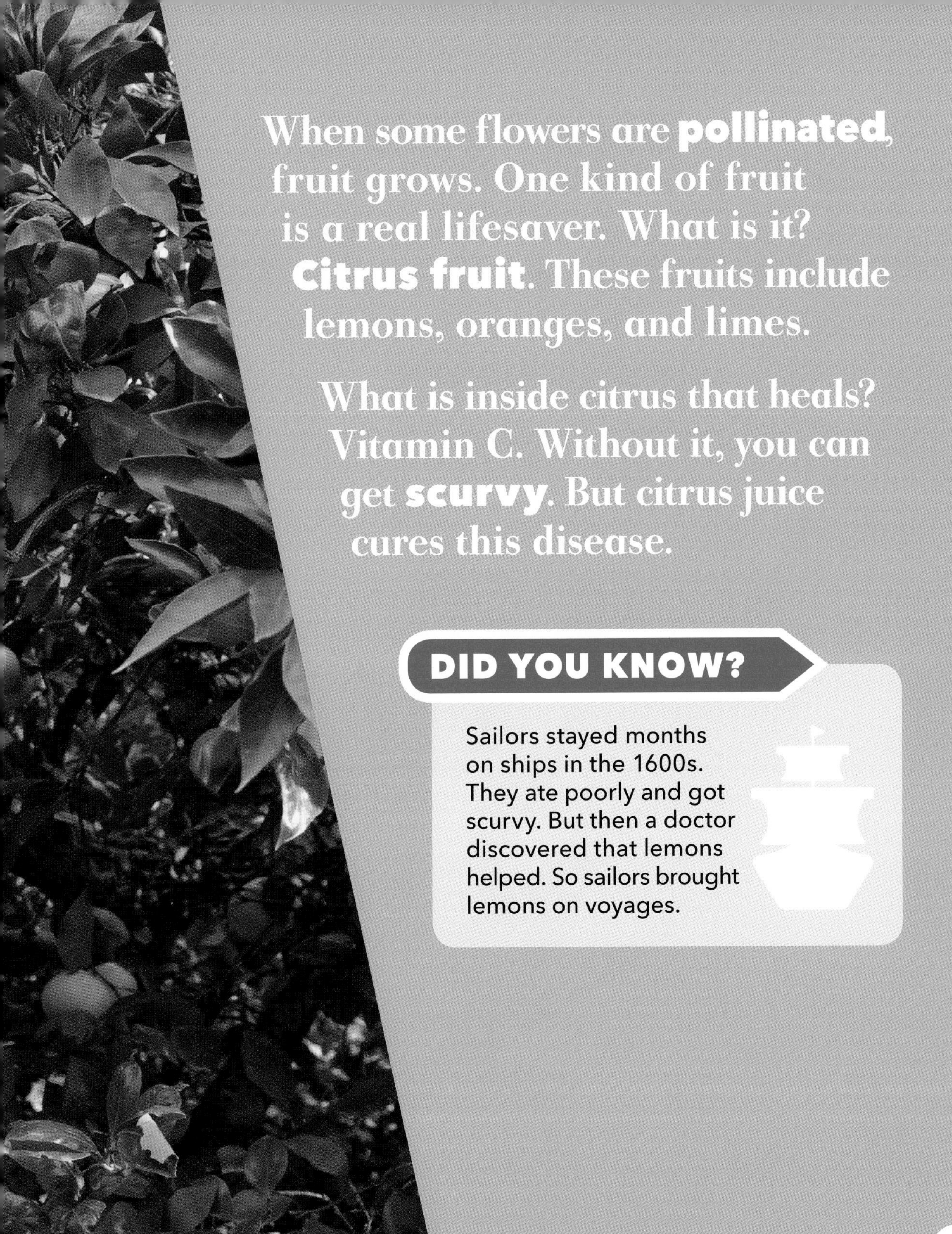

When some flowers are **pollinated**, fruit grows. One kind of fruit is a real lifesaver. What is it? **Citrus fruit**. These fruits include lemons, oranges, and limes.

What is inside citrus that heals? Vitamin C. Without it, you can get **scurvy**. But citrus juice cures this disease.

DID YOU KNOW?

Sailors stayed months on ships in the 1600s. They ate poorly and got scurvy. But then a doctor discovered that lemons helped. So sailors brought lemons on voyages.

CHAPTER 2

USEFUL LEAVES

A plant's leaves can heal, too. One plant looks like it might hurt you. Its thick leaves have prickly points. But they hide something helpful inside.

aloe plant

What is it? An aloe plant! Cut a leaf open. You'll see a thick liquid. You can rub it on a scrape or burn. It helps skin heal quickly.

tea tree

Growing in Australia is the tea tree. Its leaves have oil that kills germs. Australians used to crush its leaves and smell them when they had colds. They put the leaves on cuts to help them heal. Now many people use the oil to keep their skin clean. It can also be used to fight **head lice**.

Rub this plant's leaves with your fingers. Now smell. It is minty fresh! It's the mint plant. In a cream, it cools skin. People put it on bug bites or **rashes**. In a liquid, it helps you breathe better when you're stuffed up.

mint

CHAPTER 3

BARK AND ROOTS

Other parts of plants help people, too. This plant's bark holds its healing powers. It is a willow tree!

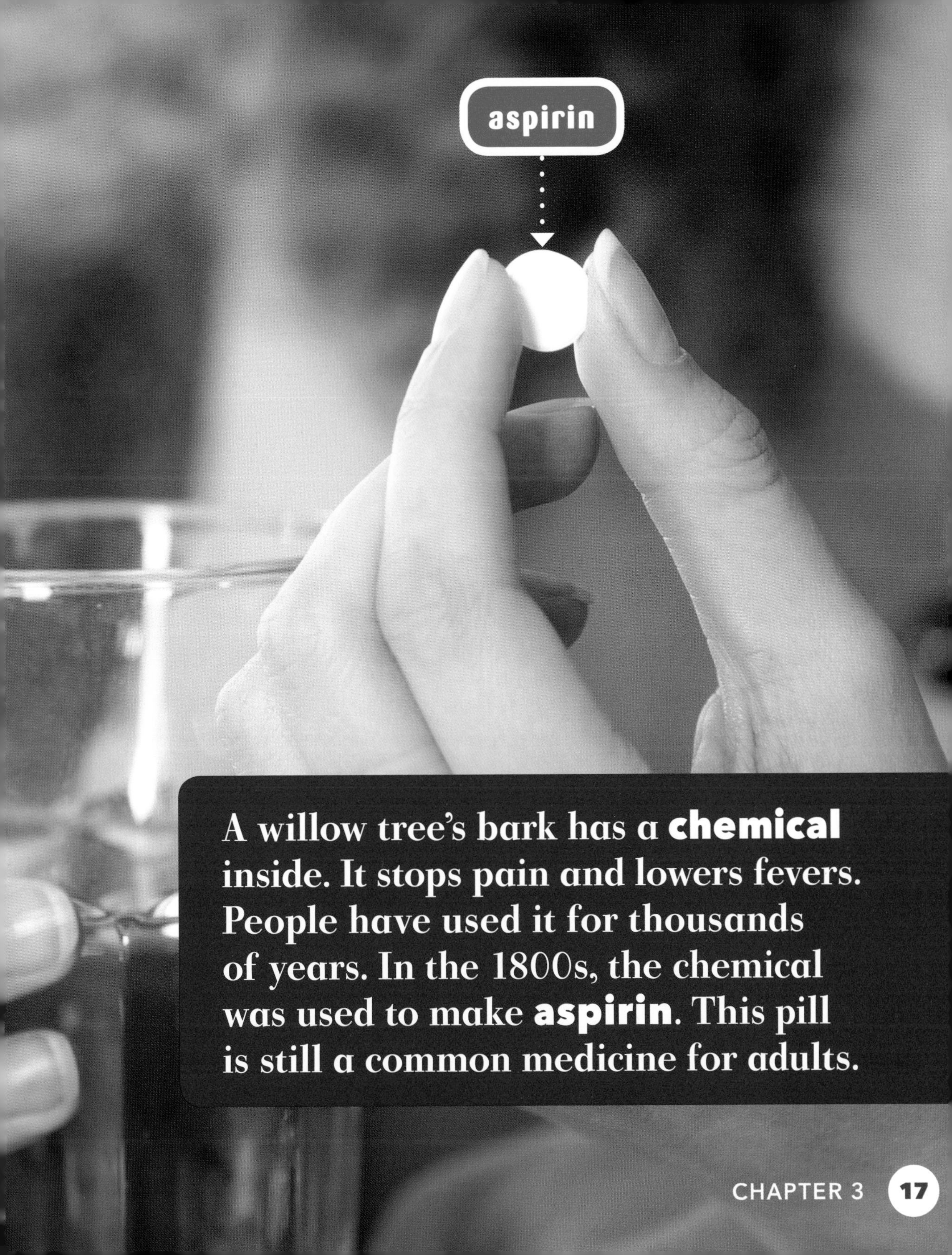

A willow tree's bark has a **chemical** inside. It stops pain and lowers fevers. People have used it for thousands of years. In the 1800s, the chemical was used to make **aspirin**. This pill is still a common medicine for adults.

Another tree grows in South America. It is an **evergreen** tree called the cinchona. Its bark has a chemical that lowers fevers. It is made into a medicine called **quinine**. People take it for **malaria**.

DID YOU KNOW?

Many plants heal. But many can also be **poisonous**. Never eat from a plant unless you know it is safe. Always ask an adult first.

Even below ground, plants have healing powers. The **roots** of ginger plants flavor food. But they also help with upset stomachs. Ginger keeps you from getting sick.

Plants have some incredible abilities. From their flowers to their roots, they heal us in many ways.

TAKE A LOOK!

Different parts of a plant can be used for medicine.

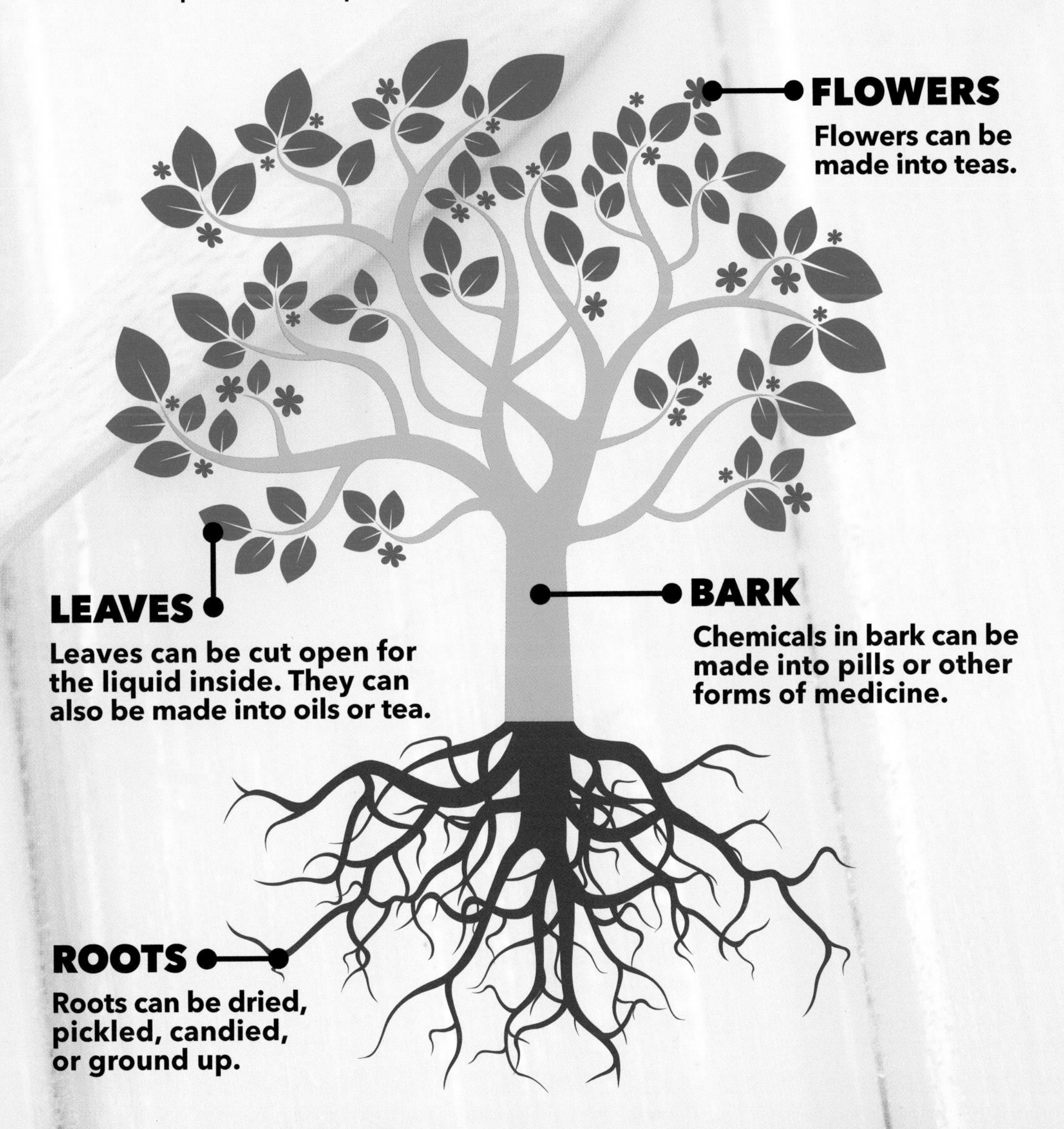

ACTIVITIES & TOOLS

TRY THIS!

SLEEPY LAVENDER MIST

The smell of lavender can help you sleep. Make this lavender mist with an adult's help. Then spray it on your pillow for a sweet night's sleep.

What You Need:

- 4 oz spray bottle
- witch hazel
- 6 to 7 drops of lavender oil

1. **Add the lavender oil to the spray bottle.**
2. **Fill the bottle with witch hazel.**
3. **Shake the bottle well to combine the oil with the witch hazel.**
4. **Mist your pillow each night with the spray. Shake the bottle before each use. Sweet dreams!**

GLOSSARY

aspirin: A drug adults can take to relieve pain and lower fevers.

chemical: A certain substance inside a plant.

citrus fruit: Fruit that is juicy and has acid, such as lemons or oranges.

evergreen: A bush or tree that keeps its green leaves all year long.

head lice: Small insects that attach themselves to people, often on their head, to feed on their blood.

malaria: A disease that people get from mosquito bites.

medicine: A substance used to treat someone who is sick.

mummies: Dead bodies preserved in a special way and wrapped in linen so that they last a very long time.

poisonous: Able to harm or kill with a harmful substance.

pollinated: When pollen from one plant is carried to another so that it can make fruit and seeds.

quinine: A medicine used to treat malaria.

rashes: Spots or red patches on the skin caused by allergies or an illness.

roots: The parts of plants that grow underground and absorb nutrients and water.

scurvy: A disease caused by a lack of vitamin C.

INDEX

TO LEARN MORE

Learning more is as easy as 1, 2, 3.

1) Go to www.factsurfer.com
2) Enter "healingplants" into the search box.
3) Click the "Surf" button to see a list of websites.

With factsurfer, finding more information is just a click away.